HIGH-FLYING ADVENTURES

A Coloring Book of Military Aircraft

This book belong to

_____

_____

Preface

Welcome to "High-Flying Adventures: A Coloring Book of Military Aircraft"! This book is a thrilling journey through the world of military aviation, featuring some of the most iconic and powerful aircraft ever built. From fighter jets and bombers to transport planes and helicopters, this coloring book has it all.

Whether you're an aviation enthusiast, a fan of military history, or simply love to color, this book is sure to provide hours of enjoyment. Each page features a different aircraft in action, ready for you to bring to life with your own creativity and imagination.

As you color in these pages, you'll learn about the history and capabilities of each aircraft, and gain a deeper appreciation for the brave men and women who fly them. You'll also have the opportunity to explore the different colors and patterns used in military aircraft design, and experiment with your own color combinations.

We hope that this coloring book inspires you to explore the exciting world of military aviation, and encourages you to learn more about these incredible aircraft and the people who operate them. So grab your colored pencils, put on your flight suit, and get ready for some high-flying adventures!

Jet Fighters

Fighter jets are some of the most advanced and capable military aircraft in the world. Designed to engage enemy aircraft and establish air superiority, these planes are equipped with powerful engines, advanced avionics, and deadly weapons systems.

Brief History

The development of fighter jets can be traced back to World War I, when aircraft were first used for combat. Over the years, these planes evolved from simple biplanes to sophisticated jets, incorporating advanced technologies such as radar, missiles, and stealth.

Engines and Capabilities

Today's fighter jets are powered by advanced jet engines, which provide the thrust needed to reach supersonic speeds and maneuver at high altitudes. Many of these engines are designed to be highly efficient, allowing the aircraft to stay airborne for extended periods of time.

In addition to their engines, fighter jets are also equipped with advanced avionics and weapons systems, including radar, missile guidance systems, and cannons. These technologies enable the aircraft to engage enemy planes from a safe distance, and to quickly detect and respond to threats in real time.

Various Types

There are many different types of fighter jets in service today, each with its own unique capabilities and design features. Some of the most well-known types include the F-16 Fighting Falcon, the F-15 Eagle, and the F-22 Raptor. These planes are designed for different missions, including air-to-air combat, ground attack, and reconnaissance.

Wars

Fighter jets have been used in many conflicts over the years, including the Korean War, the Vietnam War, and the Gulf War. These planes have proven to be highly effective in establishing air superiority, and have played a critical role in many military operations.

Stealth Bomber

One of the most advanced and secretive fighter jets in service today is the B-2 Spirit stealth bomber. This plane is designed to evade detection by radar, and is capable of delivering precision-guided bombs to targets deep inside enemy territory. With its unique shape and advanced technology, the B-2 Spirit represents the pinnacle of stealth aircraft design.

Future of Jet Fighters

Looking ahead, the future of fighter jets is likely to be defined by continued advancements in technology and design. Newer aircraft such as the F-35 Lightning II are designed to incorporate the latest avionics, weapons, and propulsion technologies, and are expected to provide even greater capabilities than their predecessors.

In conclusion, fighter jets are some of the most advanced and capable military aircraft in the world. From their origins in World War I to the present day, these planes have played a critical role in establishing air superiority and protecting national security. With ongoing advancements in technology and design, the future of fighter jets promises to be even more exciting and impressive.